P9-DHT-925

Valparaiso Public Library
103 Jefferson St.
Valparaiso, IN 46383

100 DAYS OF COOL

by Stuart J. Murphy   illustrated by John Bendall-Brunello

HarperCollins Publishers

J 513.211 MUR   VAL
Murphy, Stuart J., 1942-
100 days of cool /
33410007376389

PORTER COUNTY PUBLIC LIBRARY SYSTEM

Valparaiso Public Library
103 Jefferson St.
Valparaiso, IN 46383

MAR 0 2 2004

LEVEL
2

To Cathy Kuhns—a very cool teacher,
not just for 100 days, but all year long
—S.J.M.

To my "cool" niece, Camelia
(although she's now too old for this book!),
and, as ever, to my adorable wife, Tiziana
—J.B-B.

The publisher and author would like to thank teachers Patricia Chase, Phyllis Goldman, and Patrick Hopfensperger for their help in making the math in MathStart just right for kids.

HarperCollins®, ✦®, and MathStart® are registered trademarks of HarperCollinsPublishers. For more information about the MathStart series, write to HarperCollins Children's Books, 1350 Avenue of the Americas, New York, NY 10019, or visit our website at www.mathstartbooks.com.

Bugs incorporated in the MathStart series design were painted by Jon Buller.

100 Days of Cool
Text copyright © 2004 by Stuart J. Murphy
Illustrations copyright © 2004 by John Bendall-Brunello
Manufactured in China by South China Printing Company Ltd. All rights reserved.

Library of Congress Cataloging-in-Publication Data
Murphy, Stuart J.
100 days of cool / by Stuart J. Murphy ; illustrated by John Bendall-Brunello.
p.   cm. — (MathStart)
"Level 2."
"Numbers 1–100"
Summary: Four students arrive on the first day of school looking cool and their teacher challenges them to keep it up as they count down one hundred days to a cool celebration.
ISBN 0-06-000121-6 — ISBN 0-06-000123-2 (pbk.)
1. Counting—Juvenile literature.   [1. Counting.]   I. Title: One hundred days of cool.
II. Bendall-Brunello, John, ill.   III. Title.   IV. Series.
QA113 .M884   2003   2002019061

Typography by Elynn Cohen   1 2 3 4 5 6 7 8 9 10 ❖ First Edition

100 DAYS OF COOL

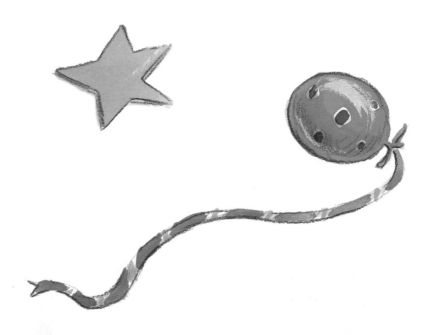

"Hey! Why do you guys look so weird?" asked Toby.

It was the first day of school. Maggie, Nathan, Yoshi, and Scott looked really wild.

4

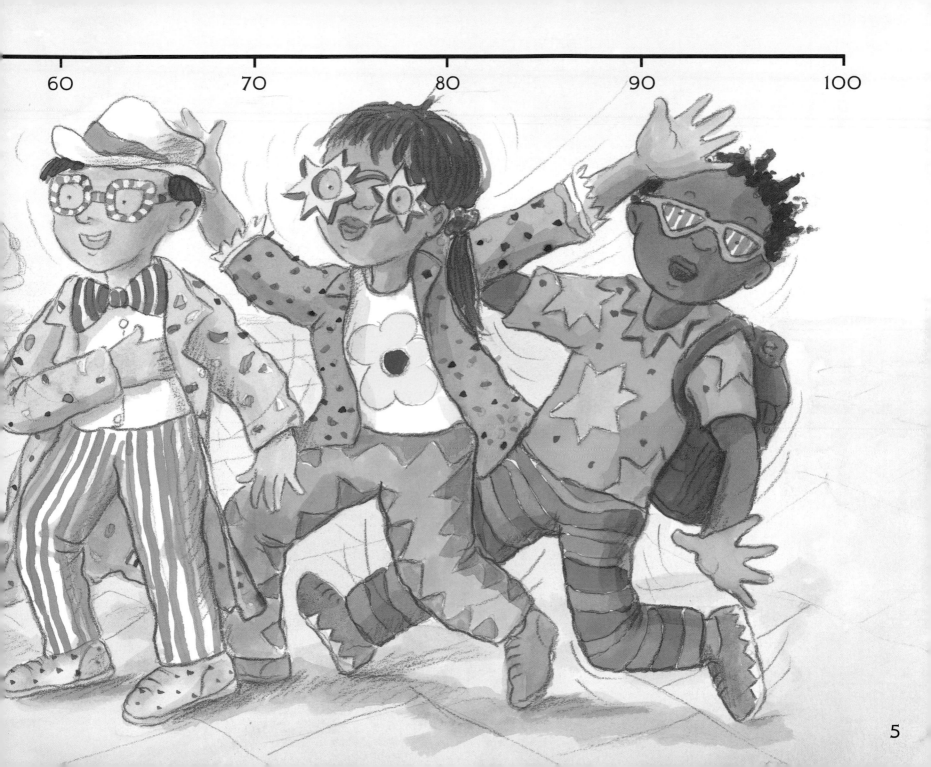

60 70 80 90 100

5

"Didn't you hear?" asked Maggie. "Our new teacher, Mrs. Lopez, is going to have us celebrate one hundred days of cool. So we dressed as cool as we could."

"Not COOL," said Toby. "SCHOOL!"

"Oh no," grumbled Scott. "Leave it to Maggie to get it wrong."

"We don't even have time to go home and change," said Yoshi.

"Well, then, let's go for it!" said Nathan.

Mrs. Lopez couldn't believe her eyes when the four cool kids walked into the classroom.

"We're all set for the first day of cool," Nathan announced.

"The first day of *cool*?" asked Mrs. Lopez. "Oh, I get it! Great idea! If you can keep this up for 99 more days, we'll have a cool party to celebrate. Do you think you can do it?"

"You bet!" shouted Maggie. The others agreed.

But the very next day, Maggie, Nathan, Scott, and Yoshi all looked normal when they arrived at school. There was not one sequin or pair of sunglasses among them.

"What happened?" Toby teased. "Did you give up already?"

"Day 2 and we're still cool," said Yoshi as they lifted up the legs of their jeans.

"Cool socks!" someone yelled from the back row.

Ha! They still have 98 days to go!

11

5 8

10

0   20   30   40   50

Days went by. The cool
kids kept going.

On Day 5 they
decorated their bikes.

On Day 8 they each
wrote their eight favorite
jokes on the blackboard.

On Day 10 they were
especially cool. They
came to school dressed
in clothes from the '70s.

13

17 21 25

0 10 20 30 40 50

Sometimes their ideas backfired.

On Day 17 they tried walking backward all day. That turned out to be sort of un-cool.

On Day 21 they wore boxer shorts over their pants. Maggie's mother almost didn't let her out of the house.

14

60   70   80   90   100

On Day 25 they dyed their hair 4 different colors.
That was cool—until they tried to wash it out.

Okay, 25 done, but a whopping
75 to go.

15

On Day 33 the cool kids pasted sparkles on their faces.
On Day 41 they announced that they were volunteering
to read books at the Oak Hills Senior Center after school.
"How cool is that?" said Mrs. Lopez.

On Day 49 they wore shirts that were half white and half black. "We're getting there!" they shouted.

They're amost halfway.

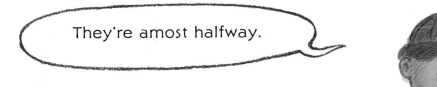

But they couldn't think of an idea for Day 50. They sat in the lunchroom talking about it.

"Come on, Maggie," said Yoshi. "You're good at ideas."

"I can't think of a thing," said Maggie.

"What about goldfish?" asked Nathan. "Goldfish are cool."

"There is nothing cool about goldfish," Scott said.

Toby was walking by. He heard them talking.

"I knew you'd never make it!" he said. "You didn't even get halfway."

Rosa was nearby too. "You can't give up now," she said. "We're almost halfway to the party! Hey, everybody! Come and help us think of ideas!"

All the kids in the class came over. Yoshi started taking notes. Soon she had a long list.

On Day 50 the kids wore scarves and mittens. They said they were cool.

On Day 75 they tried to speak Spanish all day.

On Day 82 they wore hats made out of their favorite foods. "Way cool!" said Toby, stealing a chocolate-chip cookie.

They still need almost 20 new ideas.

23

On Day 99 they each brought in a collection of 99 things.

"What are you going to do
tomorrow?" asked Rosa.
"Well, I'll tell you," said
Maggie. "No, actually, I won't.
It's a surprise."

When the cool kids got to school on Day 100, the whole class was already there.

Yoshi was wrapped up in cardboard. Scott had a plastic garbage bag around him. Maggie and Nathan were wearing their parents' raincoats. All four were covered from head to foot.

"Get ready!" said Maggie. "One . . . two . . ."

27

". . . three!"
And off came the covers.
Everybody cheered.
They had made it through 100 days of cool!

Mrs. Lopez brought out the food and the cool party got under way. But Scott didn't look happy.

"What's up, Scott?" asked Mrs. Lopez.

"What are we going to do
tomorrow?" said Scott.
"All the fun is over."

In *100 Days of Cool*, the math concept covers the numbers from 1 to 100. The number 100 is an important benchmark for children as they become familiar with place value. Many schools celebrate their 100th day as a way to culminate their study of the numbers 1 to 100.

If you would like to have more fun with the math concepts presented in *100 Days of Cool*, here are a few suggestions:

- As you read the story, point out the number line to the child. Talk about what day it is and how many more days remain until the 100th day of school.

- Make a number line similar to the one shown in the book on a long, thin sheet of paper. Fold the number line in half and in half again. Use the folds to show how day 25 is $\frac{1}{4}$ of the way to 100, day 50 is halfway, and day 75 is $\frac{3}{4}$ of the way.

- Using a cereal such as Froot Loops, make a necklace with 100 loops. Group colors by ten. (For example, you could alternate 10 orange loops with 10 yellow.) Count how many color groups the necklace has.

- Look at a calendar with the child. Starting on January 1, find the 100th day of the year. You and the child can each make a guess about the day—what month will it fall in? What day of the week will it be? Then see if you got it right. Try the same thing again, this time counting from today's date or from the child's birthday to find the 100th day of his or her new year.

Following are some activities that will help you extend the concepts presented in *100 Days of Cool* into a child's everyday life:

100 Collection: Start a collection of 100 things. For example, you might try collecting pennies, marbles, or buttons.

Domino Train: Give the child a set of dominos and have him or her try to make trains (or lines of matching dominos) with exactly 100 dots. How many trains can he or she make?

Grouping Pennies: Take 100 pennies and collect them into groups. Each group must have the same number of pennies, and the groups may not be smaller than 3 or larger than 15. How many different ways can you group the pennies so that there are none left over?

The following books include some of the same concepts that are presented in *100 Days of Cool*:

- 100TH DAY WORRIES by Margery Cuyler

- THE 100TH DAY OF SCHOOL by Angela Shelf Medearis

- 100 SCHOOL DAYS by Anne Rockwell